I·N·S·I·D·E DK G·U·I·D·E·S

ANIMAL HOMES

Written by
BARBARA TAYLOR

A snail's home is the shell on its back

A DK PUBLISHING BOOK

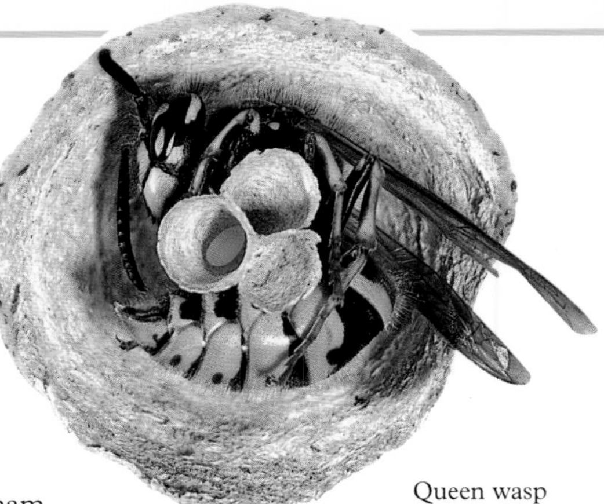

Queen wasp building a nest

Editor Melanie Halton
Senior art editors Diane Klein, Neville Graham
Managing editor Gillian Denton
Managing art editor Julia Harris
Editorial consultant David Burnie
US editor Camela Decaire
Picture research Sam Ruston
Production Charlotte Traill

Photography Andy Crawford, Geoff Brightling
Modelmakers Thurston Watson, Chris Reynolds and the team at BBC Visual Effects

First American Edition, 1996
2 4 6 8 10 9 7 5 3 1

Published in the United States by
DK Publishing, Inc.,
95 Madison Avenue,
New York, New York 10016

Blenny fish

Hermit crab in its second-hand shell home

Visit us on the World Wide Web at
http:/ /www.dk.com

Earthworm

Inside a termite tower

A CIP catalog record is available from the Library of Congress.
ISBN 0-7894-1012-5

Reproduced in Italy by G.R.B. Graphica, Verona
Printed in Singapore by Toppan

Spider's egg cocoon

Animal residents of a compost heap

House spider

Sleeping hedgehog

Contents

Young mason bee waiting to hatch

Building a home

From termite towers and beaver lodges to silkmoth cocoons and wasps' nests, many animals build amazing homes to protect themselves and their young from predators and the weather. They make use of natural materials, such as sticks, grass, and mud, as well as materials made inside their own bodies, such as silk and beeswax. Some animals cheat by living in homes built by other animals. Inside the homes animals may live in large colonies, as family groups, or alone.

Nursery rooms

The eggs and larvae of gall wasps cause plants, such as oak trees, to swell around them, forming a structure called a gall. One or more larvae live in a gall, feeding on the plant tissues. Larvae turn into pupae and then adults, which chew their way out of the gall.

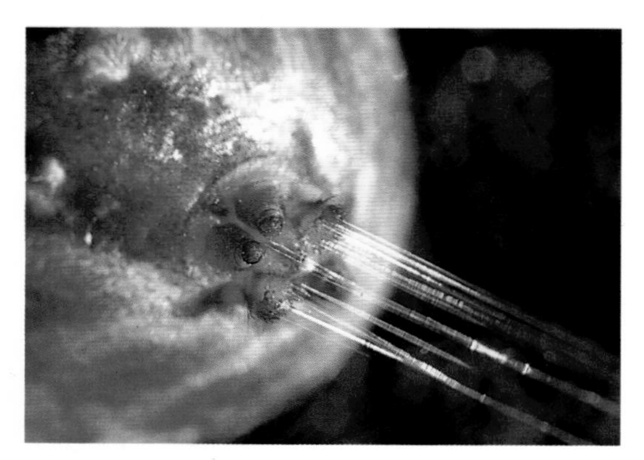

Strong silk

Silk is the strongest of all natural fibers. A spider makes silk in glands near the back of its body called spinnerets. Silk emerges as a sticky liquid and has to be pulled out of the spinnerets, usually by the legs. As the silk is stretched, it forms strong, elastic threads used for building.

Nest weavers
Harvest mice weave compact, grassy nests, and line the interiors with finely chewed grass or thistledown.

Natural resources

Wasps make building materials from natural ingredients. They mix fragments of wood with saliva to make a wet pulp that can be molded easily to build a nest. When the wood pulp dries, it forms a strong, papery material.

House share

The houses people build often provide food and shelter for many animals, from swallows under the eaves to earwigs in the basement. As more and more natural habitats are destroyed every day, animals that can live with us have a leg up on survival.

Building tools

This beaver's razor-sharp front teeth are used to fell the trees and branches it needs to build its home. The tools animals use to build their homes tend to be innate, such as teeth, claws, jaws, beaks, and feet, but the structures they construct can be very elaborate.

Digging a home

Moles are among the best diggers in the animal world. Their shovel-shaped hands and long nails are linked to strong bones and chest muscles. Other burrowing animals, such as prairie dogs, badgers, rabbits, and foxes, also use sharp claws and strong muscles for digging their homes.

Backpackers

Animals such as tortoises carry their homes on their backs. The shell is a part of the animal's body into which it retreats to avoid predators and bad weather.

Sticky larvae

Tree ants in Asia make a nest from leaves stuck together with silk from their young larvae. Some of the adults hold two edges of a leaf together with their jaws and feet while other adults hold the young larvae. Squeezing the larvae produces the silk that glues the leaves together.

Shell home
The hard, protective shell home of the tortoise is carried on its back.

The finished home

Inside a finished nest of silk and leaves, weaver ant larvae stay safe and dry. The leaves are bound by crisscrossing threads of silk into a well-camouflaged shelter. When the larvae are ready to pupate, they do not build their own cocoons. The nest is like one huge cocoon, inside of which the larvae turn into adults.

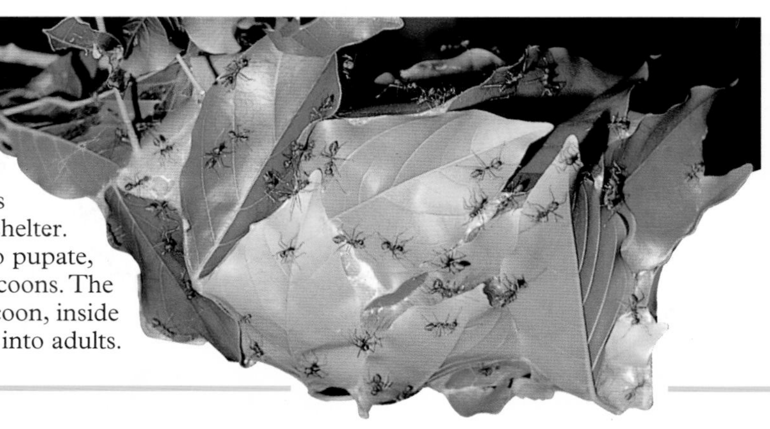

Termite tower

As tall as a giraffe, the towers of *Macrotermes* termites dominate parts of the African savanna like skyscrapers in a city. It takes the termites between 10 and 50 years to build these incredible homes, which relative to their matchstick size, are the largest structures of all living creatures. A single termite mound provides protection from predators and a stable, humid environment for millions of temites, living inside in almost total darkness. There are many shapes and sizes of termite mounds, depending on the species and the climate it inhabits.

Solid walls
The outer walls are up to 20 in (50 cm) thick and as hard as concrete.

Ventilation
Inside the mound air moves through a network of tunnels and chimneys.

Air conditioning
Wind blows over the chimneys, dispersing the warm, moist air produced by the termites and their fungus gardens (pp. 12-13). Cool air rises up to replace the lost warm air.

Termite colonies

This spectacular termite mound on the grasslands of Kenya belongs to *Macrotermes* termites. They build these huge towers to survive the hot, dry climates they inhabit. Termite colonies provide homes and food for many other animals as well, and help recycle dead and decaying material.

Umbrella mounds

In tropical rain forests, where there is always heavy rainfall, termites build mounds with several overhanging roofs. The roofs work like umbrellas, deflecting rain away from the nest and protecting the inner chambers.

Compass nests

Australian Magnetic termite towers are wedge-shaped and run north to south. The flat sides face the sun in the morning and evening, which warms the nest. At midday, very hot sun rays hit the thin edge, preventing overheating.

Treetop homes

Some termites build homes on tree branches using vegetable matter mixed with saliva and droppings. Mud barriers above the nests help funnel rain away from the nests.

Nest building

A *Macrotermes* termite tower, left, is started by the king and queen, above. They fly away from their old tower, throw off their wings, and find a place to build a new home. First they dig a hole about 1 ft (0.3 m) deep into the soil. They then hollow out a chamber in which they mate so that the queen can start laying her eggs. The termites that hatch out of the eggs take over the work of nest building. In just a few months an entire new termite city exists.

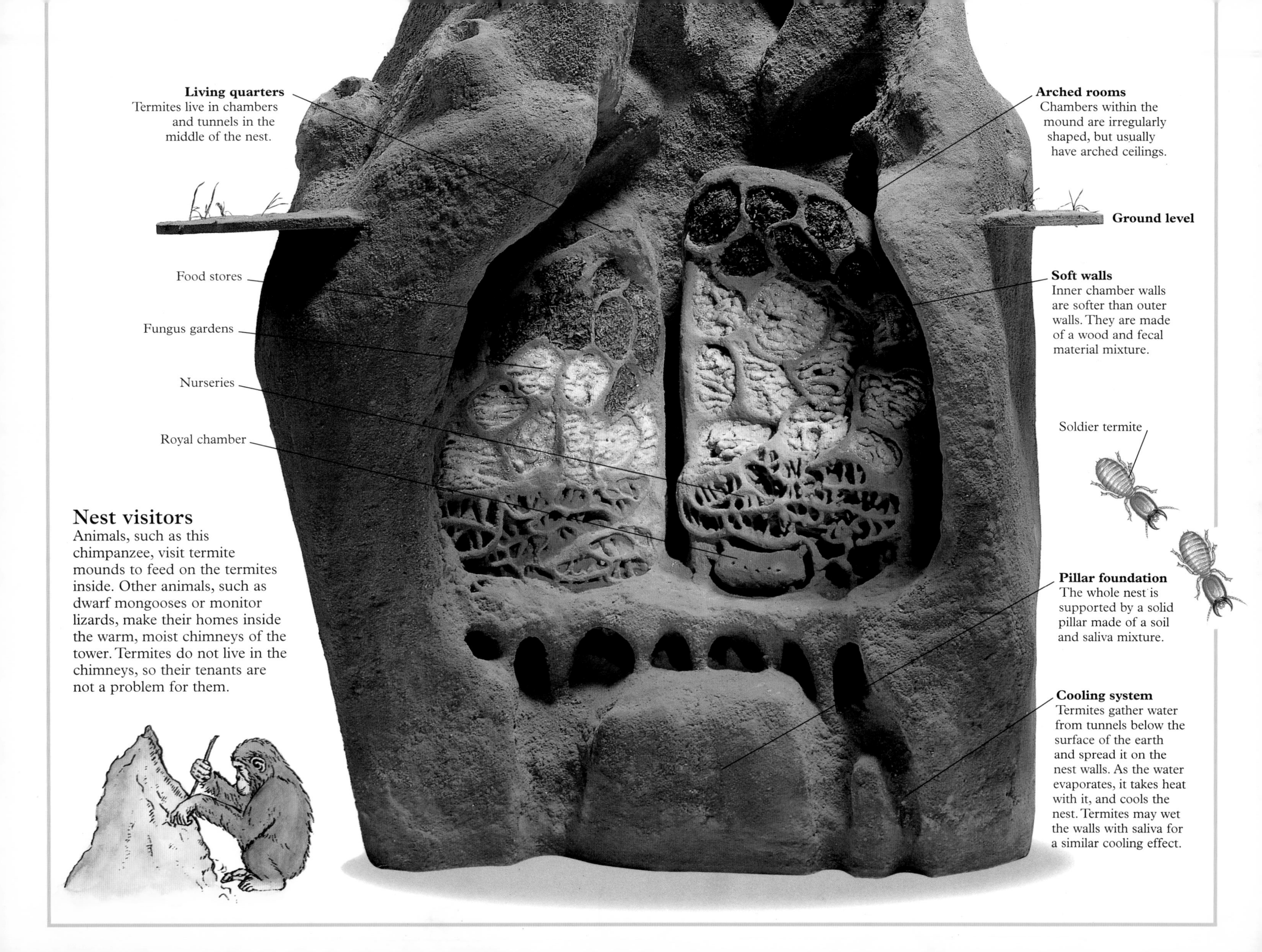

Nest visitors

Animals, such as this chimpanzee, visit termite mounds to feed on the termites inside. Other animals, such as dwarf mongooses or monitor lizards, make their homes inside the warm, moist chimneys of the tower. Termites do not live in the chimneys, so their tenants are not a problem for them.

Inside a termite tower

Zooming into the middle of a *Macrotermes* termite nest reveals fascinating details about the arrangement of the "rooms" inside. There are four main areas of activity – fungus gardens and food storage areas toward the top of the tower, and nursery areas and a royal chamber toward the bottom. The shape and size of the chambers vary, but they usually have arched roofs. Termites live most of their lives in the darkness of their mud tower, bustling from room to room along tunnels, emerging only to collect food and to mate.

A closer look
The inside of the *Macrotermes* mound, right, shows the many areas where the termites live. The nest is separated into zones, each with a different use. These zones are linked by a maze of connecting tunnels.

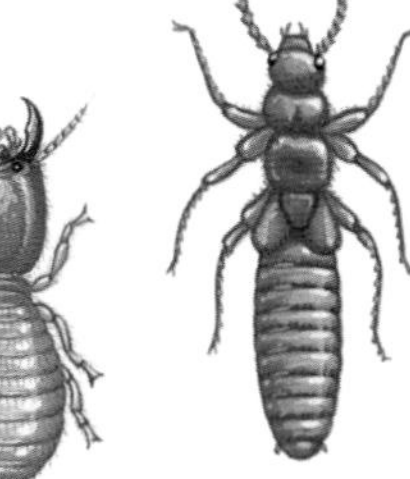

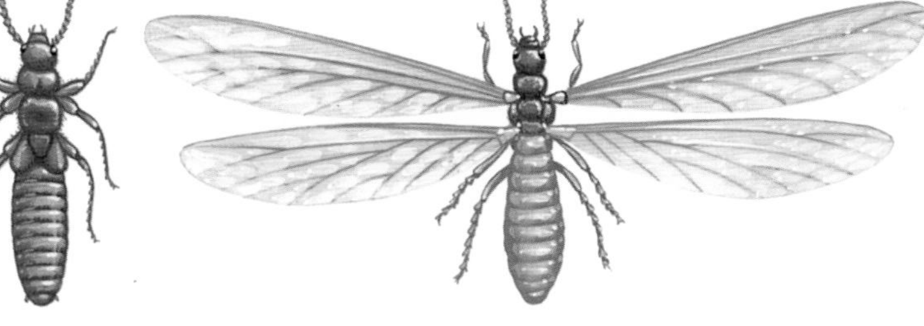

Egg **Worker** **Soldier** **King termite** **Alate**

Job sharing

There are three different types of termites within a mound, each with its own job to do. Most termites are workers who collect food and do all the work. A few are soldiers that guard the nest. Workers and soldiers live up to five years. The king and queen develop from fertile termites called alates, which hatch out once a year. The king and queen live for up to 70 years.

Queen termite

In a termite fungus garden
In this fungus "garden" a worker and soldier are munching. The blind soldiers must be fed and cared for by workers.

Silver tower
On one night each year, the winged alates emerge from the termite tower. Their thousands of shiny wings make the tower look like a silver castle. Each alate finds a mate and flies away to start a new colony.

Tasty termites
The bodies of winged alates are packed with nutrients. They are an important source of food for both people and wild animals. People in Kenya own individual mounds and trap alates as they emerge. Cooked alates taste like nuts.

Soft walls
Inner chamber walls are made of a soft, woody material that has passed through termites' guts.

Fungus chambers
Termites grow special fungus gardens. Fungi are good at breaking down vegetation, and termites use fungus to help them digest the plant material that they eat.

Nurseries
The eggs in the nursery chambers take about three weeks to hatch. Hatchlings are the size of a pinhead when they emerge from the eggs, but look like miniature adults.

Prison cell
The king and queen are sealed inside the royal chamber, where they mate frequently.

Egg factory
The queen can lay thousands of eggs in a day.

Wasp factory

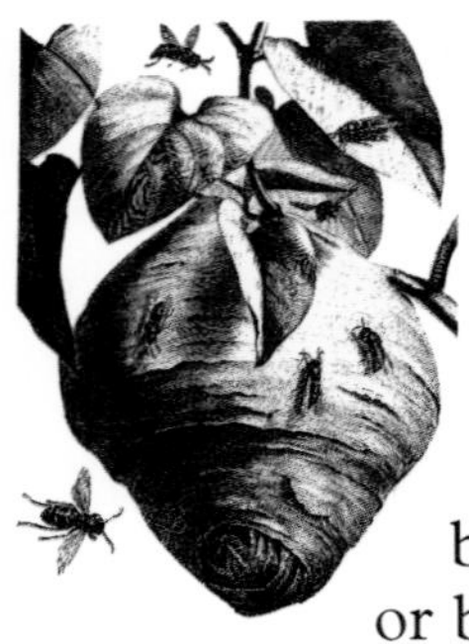

Legend has it that the Chinese learned to make paper by watching social wasps build their "paper" nests. Unlike bees, wasps cannot produce wax to build their nests. Instead, they collect particles of dry wood from posts or trees. They chew this wood into a pulp, which hardens into "wasp paper." Wasps use this paper to build nests of six-sided cells, covered in more layers of paper. The nests can be suspended from the branches of trees, the eaves of peoples' homes, or tree trunks, or be buried in holes in the soil. Each winter all the wasps living in a nest die, leaving only young queens. These queens survive the winter by hibernating and emerge the following spring, when they fly off to build a new nest and start a new colony.

Building a colony

When a young queen finds a suitable place to build her nest, she flies off to look for wood. She rasps off wood with her sharp jaws, then mixes it with saliva to make paper. The queen makes a paper stalk from which to hang the nest, adds a layer of cells, and lays an egg in each one. The eggs hatch into worker wasps, which take over the nest building, food gathering, and care of all young. At the end of summer, the workers make larger "royal cells" in which to rear new queens.

Paper walls
Protective layers of paper envelopes surround the living quarters.

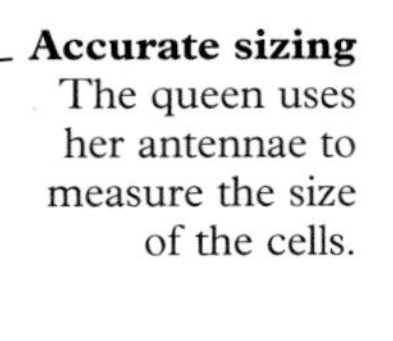

Accurate sizing
The queen uses her antennae to measure the size of the cells.

Fragile support
A stalk of paper joined to a disk is all the nest will hang from.

1 The young queen makes wasp paper and builds paper cells in which to lay her eggs. One egg is laid in each cell and has to be glued down so it will not fall out.

Sleeping queen
The queen rests before laying an egg in each of the empty cells.

Sticky cells
One egg is glued to the bottom of each cell.

Building blocks
The cells are six-sided, which makes fitting lots of cells together easy.

2 The eggs hatch into grubs, which the queen feeds with the chewed-up remains of insects she has caught.

Mature pupa
The head and legs of the developing wasp are visible.

3 When the grubs are ready to turn into pupae, they each spin a cocoon and close off the open end of their cells with doors of tough silk. About four to six weeks later, fully formed worker wasps bite their way out of the cells.

Paper home

A finished nest is roughly spherical in shape and is covered with several layers of wasp paper. It consists of eight or more layers of cells that are joined to the layers above by pillars of paper. Wasps do not store food in their cells, so they face downward with the tops closed in. This is why the queen has to glue her eggs in the cells to stop them from falling out. There is a space between the nest covering and the combs so that the wasps can move around freely.

Other wasp homes

Different species of wasps make a wide variety of nests. Social wasps make their own paper building material. Solitary wasps use wood, mud, dig holes, or make homes in the nests of other insects.

Hole for hanging nest

Narrow slit door

Clay nest
The nest of the wasp *Polybia singularis* is built largely of mud. It is so heavy it must be hung from a strong branch.

Home repair
Holes in the nest are repaired with new wasp paper.

Papery spines

Oval door

Spiked home
Chewed plant fiber is covered with hard papery spines to make the nest of the South American *Polybia scutellaris* wasp.

Stripy paper
Colored layers occur because wood is gathered from different sources.

Making space
Old paper walls inside the nest are chewed away to make room for more cells.

Doorway
Small entrance is easy to guard and regulates the temperature of the nest.

Mud nests

Mud is an ideal building material because when it is wet, it can be squashed and molded into all sorts of shapes. Many birds, such as swallows, martins, and flamingos, build their homes out of mud. The rufous hornero is nicknamed "ovenbird" after the mud nest it builds in the shape of a traditional clay oven. People also build homes from dried mud or mud bricks. When the mud dries, it sets as hard as stone, protecting the birds from predators and weather, giving the eggs and young a safe home.

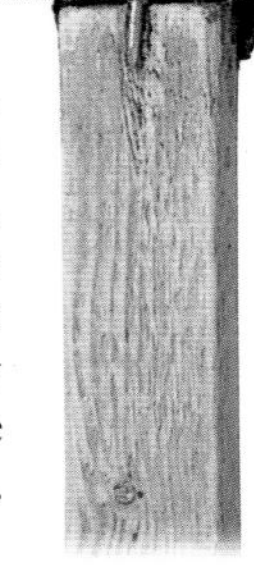

Nests in the open
Ovenbirds live on open grassland or farmland, and in parks or in cities where there are few trees to nest in. They usually build their nests on fence posts or telegraph poles.

Mud pockets
Swallows build their cup-shaped nests high on the walls of buildings or under bridges. The mud is mixed with some straw and lined with feathers to make a warm, dry pocket for eggs and young.

Teamwork
Male and female ovenbirds build a strong nest together.

Building the nest
Ovenbirds help each other build a nest, which is about the size of a football. They collect thousands of lumps of wet mud and cow dung and add dry grasses to make the nest strong and to prevent it from cracking.

Mud mounds
Flamingos build their nests by lakes and rivers. They use their bills to make mud heaps, which dry in the sun. The nests protect their eggs and young from floodwaters and the intense heat of the ground.

Baked eggs
The warm mud nest acts as an incubator for the eggs inside.

Drying bricks
For thousands of years, people have used mud to build their homes. These mud bricks were shaped in a wooden mold in Peru. The heat of the sun dries the bricks, creating a hard and versatile building material.

Nest recycling

It takes 18 days or more for the rufous hornero ovenbird to complete its mud nest, left. New nests are built every year and may be positioned on top of old nests. Sometimes old nests are taken over by other birds.

Grassy nests

Many birds build homes out of dried grass, but the most spectacular grassy nest is that of the sociable weaver bird of southwestern Africa. This bird uses a nest-building technique very similar to thatching. Carrying one straw at a time, it pushes together an enormous nest. The nest is constantly repaired and rebuilt and may be used for many years. Other weaver birds loop, twist, and knot grass stems to make woven hanging nests.

Knotted weaver bird nest

Weaving techniques

Most weaver birds use a variety of "stitches" and knots, from slip knots and half hitches to loop tucks and spiral coils. A male weaver starts a nest by building a ring of grass attached to a branch. He then perches on this "swing" while he adds a roof and an entrance tunnel. A female will line the nest with soft grass tops and feathers.

Slip knot

Half-hitch knot

Reversed winding

Spiral coil

Barbed gateways
Sharp grass stems make a fence that helps to keep out predators.

Apartment tower

Covering an acacia tree in Namibia, this huge thatched nest is home to about 300 sociable weaver birds. The massive nest of the sociable weaver bird may be 25 ft (7.5 m) long, 15 ft (4.5 m) wide, and 4 ft (1.5 m) high. Sociable weavers live and sleep in their nest all year round, and raise their young in it.

Sociable weaver bird

Safety net

At the entrance to the round nest chamber, the birds construct a threshold of short, crisscrossed grass stems. This prevents eggs and young from rolling out.

Woolly purse

Penduline tits are named after their incredible hanging nests. Made from lichens, grass, leaves, and moss, these strong nests look like woolly purses.

Thatched cottages

Sociable weavers do not build their nests by weaving or knotting. Instead, they push straw or grass into the nests in the same way a human thatcher builds a roof of dry reeds.

Thick thatch
The roof is about 12–24 in (30–60 cm) thick and is built mainly by male birds.

Sloping roofs
The roof slopes down. This helps rain drain away, keeping the nests below dry.

Home additions
As the nest grows, new chambers are added at the sides.

Nest squatters
Other birds, such as the pygmy falcon or the pied barbet, may take over empty nest chambers.

Upside-down doors
The entrances to the nest are vertical tunnels that point downward.

Mansion quarters

The huge thatched nest of the sociable weaver bird, above, has individual nest chambers underneath it where the birds live in pairs.

Cozy cocoons

Some small animals make the building materials for their homes inside their own bodies. Silk is produced from the salivary glands of insects and the abdominal glands of spiders. The delicate silk threads are used to weave shelters. These silk shelters, called cocoons, usually protect the eggs and young while they develop into adults. Some cocoons, such as those of the silkworm moth, are made entirely of silk threads, while other creatures add camouflage to their silk homes.

1 For thousands of years, silkmoths have been reared on farms. Females lay about 500 yellow eggs the size of pinheads. The eggs have a gluelike covering and stick to everything they touch.

2 Silkmoth caterpillars, called silkworms, hatch out of the eggs. They are put on feeding tables with piles of mulberry leaves. They have very weak legs and cannot walk far to feed. Silkworms grow ever bigger as they eat continuously.

3 When the silkworms are ready to pupate, they are put onto wooden racks. There they spin a long thread of silk, produced near the head. Silkworms toss their heads to wrap the silk around them, forming a cocoon.

4 As adult silkmoths hatch, they break the silk threads. To protect the silk, farmers kill the pupae before they mature. The cocoons are then sent to a special factory where the silk threads are put onto spools.

Silk threads
Each cocoon is made up of one piece of silk up to 1 mile (1.6 km) long. It takes 110 cocoons to make a silk tie, and 630 to make a silk blouse.

Head spinners
While a silkworm is spinning its cocoon, it shakes its head about 300,000 times.

Greedy worms
Silkworms keep feeding until they each weigh over 10,000 times as much as they did when they hatched from the egg.

Inside a cocoon

A silkmoth pupa needs the protection of its cocoon because it cannot move or defend itself. Inside the pupa, the silkworm's body is broken down and rebuilt into the body of an adult moth. This amazing transformation is called metamorphosis.

Earthworm cocoon

The lemon-shaped cocoons made by earthworms for their eggs are very tough. They are produced from a thick band on a worm's body, the clitellum. About 12 eggs are placed in the cocoon, but only one worm usually hatches.

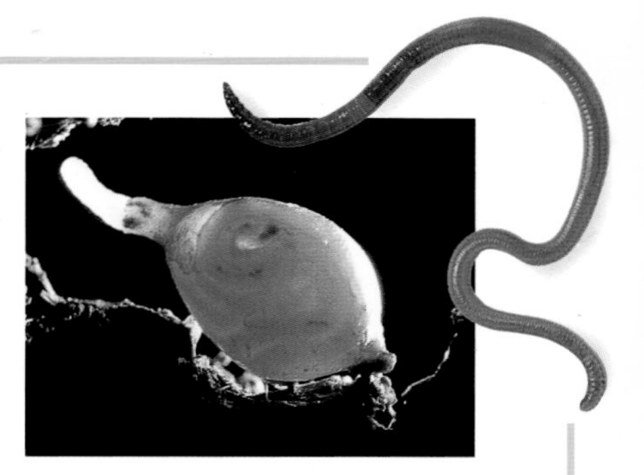

Waterproof jacket
The silk cocoon makes a hard, waterproof shell around the pupa.

Leafy shelter

The caterpillar of the oak silkmoth spins its cocoon among oak leaves, binding the leaves with silk. Wild silkmoths do not spin silk threads as long or as strong as those of domesticated silkmoths.

Pupating worm
A silkworm takes about three weeks to change into an adult.

Silk-wrapped ants

The developing pupae of yellow meadow ants are encased inside silk cocoons of varying sizes. These cocoons are often mistakenly called "ants' eggs."

Molting moth
Because its skin cannot stretch to fit its growing body, a silkworm sheds its skin about four times while it is a caterpillar.

Walking home

Caddis fly larvae live underwater in long, thin cases that they pull around with them when they move. The shapes of the cases vary per species and are built of different materials, such as sand, sticks, leaves, or shells. These materials are stuck together with silk from the larva's mouth.

Rotting log

Living trees provide homes for a great variety of birds, mammals, and insects, but once a tree dies or a branch falls to the ground, new tenants move in. Bark beetles and fungi feed under the bark, rotting and softening the wood. Other insects then bore into the wood, leaving pathways for bacteria and more fungi. Hundreds of animals live in a rotting log, all year round, because it is warm, moist, and safe from most predators. After several years, a fallen log will crumble away and the animals will find a new home.

Scattered homes

If all the fallen trees and branches are cleared away from a forest, as many as one fifth of all the animals living there will lose their homes.

Cocoon may contain over a thousand eggs

Bark of rotting log

Wooden nursery

Some spiders attach cocoons to the inside of the bark of rotting logs. There the eggs are hidden from enemies and protected from weather. The cocoon itself also protects the eggs by trapping air and keeping the eggs warm and humid.

Plant life
Rotting logs soak up water, making them ideal habitats for plants like mosses and ferns.

Recycling the log

Living wood has chemical defenses to protect it from attack by fungi like this sulfur tuft. Once the wood dies, these defenses weaken, and fungi start feeding on the wood, causing it to rot. Bacteria feed on the wood, too, but fungi are more important in the decay process.

Toad in the hole
There may be shelter under a log for larger animals, such as toads. Toads absorb moisture through their skin, so they like the damp conditions.

Damp housing
Millipedes, too, like the dark, moist habitat in or under a log. They soon die if their surroundings dry out. Millipedes feed mainly on dead and rotting leaves, but will also eat dead worms, insects, and other mini-beasts.

Cardinal lives

Beetles, such as this cardinal beetle, have four stages in their life cycle. After mating, a female lays her oval eggs in a rotting log. The long, flattened larvae that hatch have two prongs at the tail end. The larvae take about three years to pass through a pupal stage and grow into adults.

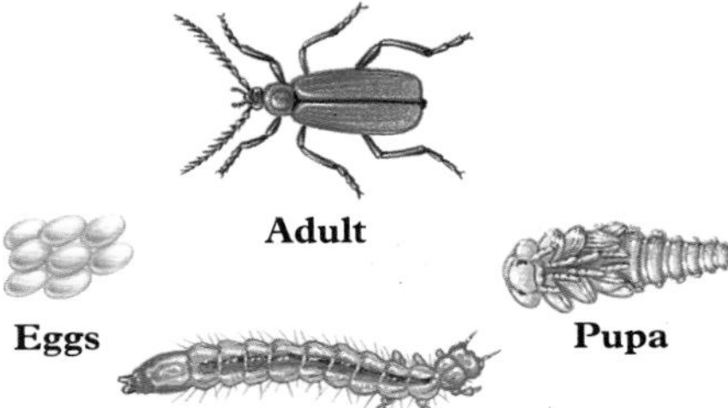

Egg-laying tube

Cardinal beetle larvae

Drilling holes

The female ichneumon fly drills a hole into a log using the long egg-laying tube at the end of her body. This fly has found a wood wasp larva and will lay her egg on it. When the egg hatches, it will eat the wood wasp larva as it grows into an adult.

Stag beetle larva

Woodcutters

Stag beetle larvae use their sharp, cutting jaws to chew their way through the wood of rotting logs. They may take as long as five years to develop into adults.

Adult stag beetle

Wood-boring beetle larva

Burrows of developing larvae

Tunnel homes

Wood-boring beetles burrow into rotting logs to feed on the wood and to make tunnel homes. The wood-borers are mostly larvae that develop into adults inside the wood. They develop slowly because wood is hard to digest.

Leaf litter

Predatory mite

Springtail

Leaf litter

The dead and rotting leaves under a fallen log are home to many tiny animals, from springtails, mites, and false scorpions to beetles, slugs, and snails. They eat either the dead leaves or each other. As the leaves are eaten away, they gradually break down and crumble.

Hunting lodge
Centipedes are attracted to rotting logs by all the possible food sources. Centipedes are hunters, killing their prey, such as worms and spiders, by injecting them with poison.

Nesting place
Bank voles sometimes build their grassy nests in tree stumps or under fallen logs. There the young are hidden from their many predators, such as foxes, and are protected from weather.

Compost heap

The warm, moist, environment of a compost heap provides an ideal site for many animal homes. The heap cannot be a permanent home though, since the rotting vegetation will eventually crumble away, leaving only a rich soil fertilizer. But while this change takes place, the rotting plants provide food for many small animals, such as worms, millipedes, slugs, and earwigs.

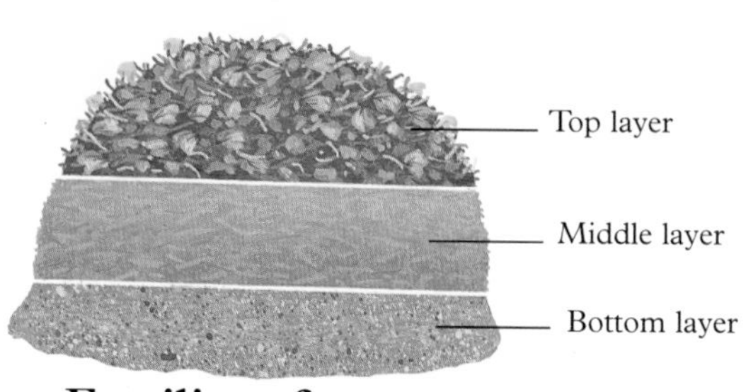

Fertilizer factory

A compost heap is much more than a pile of garbage. It is like a fertilizer factory working around the clock to make nutrients for a garden.

Hibernators

Hedgehogs visit a compost heap to feed on slugs and snails. They may hibernate in the compost over winter because the temperature in the middle of the heap stays above freezing.

Natural recycling

Compost heaps are a useful way of turning vegetation into rich fertilizer. This can then be used to enrich the soil in a garden. Many compost dwellers are specially adapted to life in a heap. They will die when put on the garden, giving the nutrients in their own bodies to the soil as well.

Winter sleep
The warmth of the heap helps keep a hibernating dormouse warm over the cold winter months.

Shiny eggs
The small, shiny eggs of snails are kept moist and warm in the heap.

Slimy slug
Without shells, slugs are in danger of drying out, so the damp compost is an ideal home.

1 A compost heap begins life as a pile of waste vegetation. Bacteria start to break down the vegetation and the heap begins to heat up. Fungi then start to grow, feeding on the tougher, woody materials, and thousands of tiny animals, such as mites and springtails, arrive to live in the heap.

2 As the bacteria, fungi, and small animals feed on the heap, they make the vegetation rot away. In the middle of the heap, eggs and young are kept warm and are hidden from predators. A heap heats up in about a day, and within a week may even be steaming.

Reptile eggs
Grass snake eggs need a warm place in which to develop.

Compost predators
Dung flies in a compost heap are fierce hunters of other insects, such as bluebottles.

Blowfly pupae

Baby blowflies
The warm compost heap is ideal for developing blowfly larvae.

Rotting mushrooms

Unlike plants, fungi cannot produce their own food. Instead, they produce chemicals that change decaying plants and animals into easily digestible food. This aids the rotting process, further breaking down the compost. The fungi produce mushrooms for reproduction as the food supply runs out.

Greedy worms
Earthworms eat their own weight of decaying plant material every day.

Sow bugs
By feeding on dead compost matter, sow bugs help speed up the rotting process.

Earwig

3 Mature compost is dark and crumbly and takes up to six months to form. Animals in the heap need air and water to survive, so a heap will not properly rot if it gets too dry, too waterlogged, or does not have enough air circulating through it.

Underground homes

Many small mammals and a few birds burrow underground to find safety, shelter, and a place to raise their young. Moles spend most of their lives underground, only surfacing to collect nest material. Most burrowing animals, however, come to the surface often to feed, gather nest materials, travel to new areas, or look for a mate. Underground homes are common in grasslands, where there are few trees to provide shelter and few tree roots to obstruct burrows. Prairie dogs, gerbils, and rabbits all live in grasslands. Other burrowers, such as foxes and badgers, prefer woodland habitats.

Hilly homes

Moles live in many habitats, such as woods, meadows, and gardens. They live underground, but can be detected by the hills that they push up as they dig soil out of their tunnels.

Tunnel diggers

The size of a mole's home depends on the richness of the soil. In places where food supplies are scarce, there are likely to be many mole hills because the mole is forced to dig more burrows and push up more hills to find enough to eat.

Food supply
A supply of fresh worms is always kept in a "pantry."

Fortress protection
A mole's grassy nest is built under a large, more permanent mole hill called a fortress.

Breeding nest
In spring the female mole breeds and gives birth in a nest lined with grass, leaves, and other soft material.

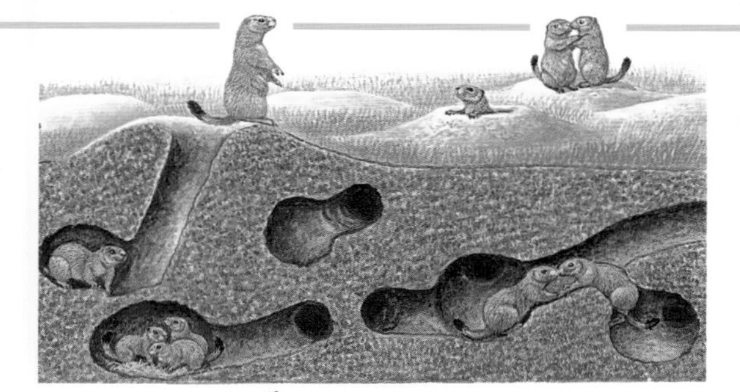

Subterranean towns

North American Prairie dogs live in underground burrows called towns. As many as a thousand individuals may live in a town. Within a town are coteries, or family units, made up of about 30 prairie dogs. Family members help defend against predators and tend the grass.

Burrowing bird

One of the few birds to live underground is the burrowing owl of the Americas. These owls often take over burrows abandoned by prairie dogs or other mammals. They will also dig their own burrows in sandy soil, using their long legs, sharp claws, and pointed bills to push the soil aside.

Sunken corridors
Tunnels may run just below the surface of the soil or be as far as 3 ft (1 m) down.

Cave homes

The people of Coober Pedy, Australia, live in underground homes built into rock. These cavelike homes protect the inhabitants from the severe and constant heat above ground.

Nest building
A mother mole must venture above ground, usually at night, to gather materials for lining her nest.

Mole patrol
As the mole bustles along its tunnels, it gobbles any worms, beetles, and insect larvae that fall in through the walls.

Home intruder
Moles live alone, so this intruder will not be welcome. In spring, however, this mole may become a mate.

Hunting halls
Tunnels used for hunting branch off from the main nest.

Spider sanctuaries

The silk-lined burrow of a trapdoor spider provides a well-constructed retreat from heat, cold, and rain, as well as an ideal place from which to ambush prey. It also helps the spider avoid predators, such as scorpions, large centipedes, and the large hunting wasps that sting trapdoor spiders and use them as living food for their young. Burrows can be simple tubes or elaborate structures with side passages, hidden doorways, and escape tunnels. The trapdoors are made of silk and soil and vary in thickness. They have silk hinges along one side that allow the doors to fall into place under their own weight. Some trapdoor spiders hold the door shut from below to keep predators out. Most trapdoor spiders spend their long lives inside their burrows, but others emerge to hunt.

Spider silk
In addition to using silk for building purposes, spiders use it to make webs for catching prey. Silk is very strong and stretchy, and webs are sticky to stop insects from escaping. The more an insect struggles, the more tangled it becomes.

Tree-trunk home
Some spiders build silken tubes on rough tree trunks. They weave pieces of bark into the tough tubes for camouflage. Others build silk homes on cliffs and rock faces and hide them with plants and other debris.

Sticky silk
Spiders produce silk, as a sticky liquid, from tubes on the abdomen called spinnerets.

Open house
Some trapdoor spiders lurk in their doorways at night, ready to snatch passing insects with their front legs. They keep their back legs inside the burrow in case of danger. They may leave silk tripwires or twigs at the door to alert them to a possible meal.

Burrowing spider
Trapdoor spiders dig out burrows using the strong spines on their fangs, which work like rakes to sweep away the soil. The spider carries little heaps of soil out of the burrow and drops them a short distance away.

Hairy legs
Special hairs on the legs are sensitive to vibrations made by prey.

Digging tools
Front legs and mouthparts help dig burrows.

Scraping fangs
Spines on the spider's fangs scrape out soil.

Hidden doorway
Silk and soil form a hinged door that fits tightly into the nest opening. The closed door hides the burrow and prevents it from flooding.

Emergency exit
A side tunnel is an escape route in times of danger.

Secret passage
To escape a predator, this Australian spider hides in a secret passage. It pulls a trapdoor shut behind it, sealing off its hiding place.

False doors
Halfway down this Australian trapdoor spider's burrow is a collapsible silk sock, piled with garbage. If in danger, the spider hides below the sock.

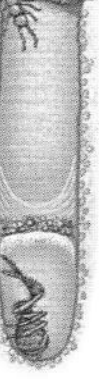

Bottoms up
This spider has an armor-plated bottom. If it senses danger, it slides down its narrowing burrow until the shield is wedged in place.

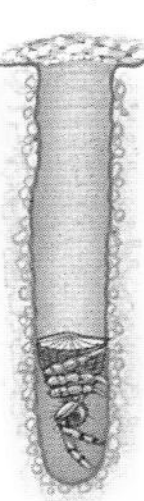

Tunnel home
The silk-lined burrow is up to 12 in (30 cm) deep and 1.2 in (3 cm) wide.

Silky eggs
A female trapdoor spider may dig a chamber in which to hang her silk egg sac. The sac may contain up to 300 eggs.

Home alone
A trapdoor spider lives alone in its burrow, and may emerge at night to hunt for food.

Under the sand

Buried beneath sandy shores, many small animals are protected from predators, waves, winds, and rapid changes in temperature. A few inches below the surface, the temperature and salinity of the water hardly vary, and the sand stays moist. Some animals live in the sand all the time, but a few, such as the green turtle, only visit the shore to lay eggs.

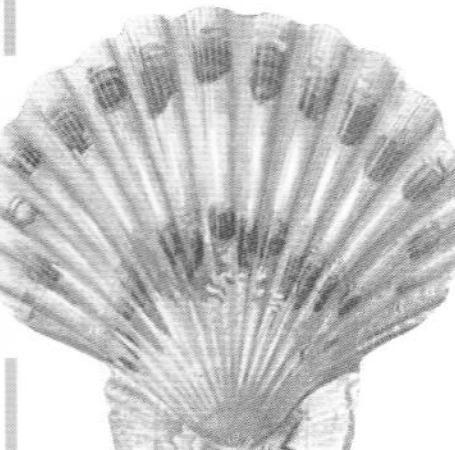

Great scallops make their homes by nestling into the seabed.

Peacock worms
When under seawater, peacock worms push out a fan of tentacles to feed. They live in tubes made of tiny pieces of mud and sand stuck together with slimy mucus.

Netted dogwhelk
This whelk can plow through sand with only its siphon sticking above the surface.

Sand dollar
The tiny spines and flat shape of the sand dollar enable it to burrow easily into sand.

Sand mason worm
The sandy tube home of this worm may be up to 10 in (25 cm) long. Only a small part sticks above the surface of the sand.

Sand gaper
This shellfish digs deep into the sand, feeding and breathing through its two long siphons.

Lugworm
The home of the lugworm can be detected by coils of waste that appear at one end of its U-shaped burrow.

Sea potato
This creature digs into the sand with large, broad spines below its body. Suckered feet open a tunnel to the surface.

Masked crab
By day, this crab lives under the sand. It breathes through a tube formed by zipping together its two antennae.

Wriggly worm
Ragworms live in burrows under the sand, or wriggle around hunting for small animals.

Edible cockle
To feed, cockles need water, so they live on the lower shore.

Sandy nest

Female green turtles lay their eggs in holes on warm, moist beaches. They cover the eggs with sand and leave them to develop and hatch. On hatching, baby turtles head for the safety of the sea to escape hungry sea birds and other predators.

Hatching green turtle

Feet first

A razor shell bores into sand with its muscular foot faster than a person can dig. It pushes down and then expands the tip of its foot to make a flat disk. This acts as an anchor, and the muscles of the rest of the foot pull the shell into the sand.

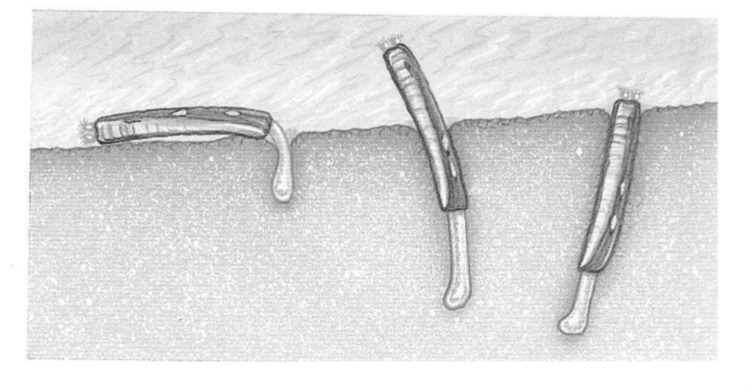

Necklace collar

Necklace shell

Shell eat shell

The necklace shell hunts shellfish such as thin tellins. It bores a neat hole through the shell of its victim with its rasping tongue and an acid it produces, to reach the fleshy body inside. The necklace shell is named after the necklace of eggs, jelly, and sand that it produces.

Burrowing fish

Lesser sand eels use their pointed lower jaws to burrow under the sand, sometimes in huge numbers. They feed on worms, plankton, small crustaceans, and fish, and are an important part of the diet of other fish and sea birds.

Starfish
Starfish live on the seabed, burrowing into the sand with its flat arms and pointed feet.

Ribbon worm
This burrowing worm traps prey with its proboscis.

Razor shell
The long, streamlined shells of the razor allow it to cut through the sand quickly.

Sea mouse
This mouselike creature is actually a worm that lives in muddy sand under the low tide mark.

Tellins
The slim, smooth shells of tellins allow them to bore easily into sand. The long siphon sucks in food, while the short one expels waste.

Crawling cucumber

Some sea cucumbers burrow below sandy shores by contracting body muscles. Tentacles around their mouths push away sand. Some sea cucumbers just move through the sand, while others dig a U-shaped burrow to live in. This exotic sea cucumber lives in the Australian shore.

Stick homes

The master builder of stick homes is the beaver, although other mammals, such as people and pack rats, also make use of this strong building material, which is readily available in many habitats. Beavers design and build their homes, called lodges, with great skill. They also change the landscape by damming rivers to create ponds around their lodges. The ponds work rather like moats around castles, making it harder for predators, such as wolves, bears, and coyotes, to reach the beavers. They also help the beavers swim close to their food sources and float heavy building materials to the lodge and the dam.

Beaver habitats
In addition to providing food and homes for a wealth of other wildlife, beaver ponds help prevent rivers from flooding.

Dam builder

Beavers know instinctively how to build dams by weaving sticks and branches together and filling in the gaps with mud and stones. Any breaks in a dam must be repaired quickly or the rushing water will tear it apart. Some dams are more than 328 ft (100 m) long and as tall as a person.

Lodge

Dam

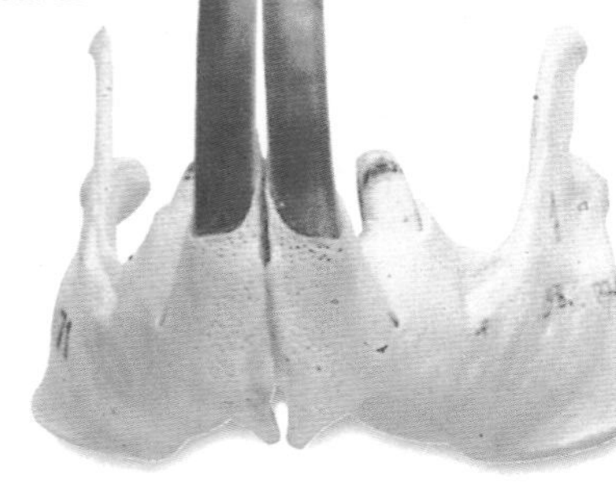

Lower jaw of a beaver

Chisel teeth

A beaver's gnawing power comes from four strong front teeth. These have hard enamel on the front, but are softer at the back. The backs of the teeth wear away more easily to make a sloping cutting edge, like a chisel. A beaver's teeth grow all the time. Constant use sharpens them and stops them from growing too long.

Lodge walls
Solid mud walls keep the beavers warm and predators out.

Underwater refrigerator
In winter, when the pond freezes, beavers eat bark and leaves from an underwater supply of branches.

Swimming beaver
Webbed back feet push a beaver through the water. Its broad, paddle-shaped tail acts as a rudder.

Floating homes

People living on Lake Titicaca, South America, build homes on floating mats of dried reeds. These strong reeds grow in shallow, marshy parts of the lake. Some lake homes are built on tall stilts to keep the living quarters dry. Just like beavers, many people around the world live on lakes for safety reasons.

Pack rat nest

Pack rat

Mounds of sticks

Pack rats, also called wood rats, build bulky nests of twigs or cacti in piles of rocks or at the base of a tree. The nest may be more than 3 ft (1 m) across and those made of spiky pieces of cactus make it impossible for predators to enter. Pack rats are attracted by bright or shiny objects and often build these into their nests.

Air conditioning
A vent at the top of the lodge lets fresh air in and stale air out.

Baby beavers
Young beavers, or kits, are born with a full coat of fur and can swim within a few hours. After about two years, they leave home to build their own lodges.

Raised rooms
The floor of a lodge is about 6 in (15 cm) above water level. There may be two levels, one for feeding, the other for sleeping.

Home repairs
Beavers use their small, agile front feet to weave each twig carefully into place.

Lumberjacks

A beaver takes only about ten minutes to fell a small tree. All the gnawing is usually done on one side of a tree, which will normally fall downhill, toward the pond.

Tunnel entrance
The underwater entrance is hidden from predators and deep enough to avoid it freezing up in cold weather.

Tide pool refuge

Many animals seek refuge in tide pools along shorelines between tides. Some animals live in one pool all their life, while others move between pools at high tide. In small tide pools, the temperature and saltiness of the water change with the weather and time of day. Larger tide pools, however, are more stable, and shelter a miniature ecosystem, from seaweeds that make their own food to plant eaters such as limpets, meat eaters such as starfish, and scavengers such as crabs.

Tide pool

Oxygen donors
During the day, seaweeds release oxygen into the water as they make their own food.

Closed anemones
When anemones are exposed above water, they retract their tentacles to stop themselves from drying out.

Creeping starfish
Starfish crawl into tide pools to find damp places to shelter, particularly at low tide.

Submerged crab
Velvet swimming crabs hide underwater to stop their bodies from drying out.

Meat eaters
Anemones look like plants, but are really carnivorous animals.

Nesting nearby

Oystercatchers often visit tide pools to feed on mussels and other shellfish. The bird uses its strong bill like a chisel or a hammer to pry or smash open the shells.

Blenny fish

Tide-pool fish

These fish make their homes by wriggling their bodies to push aside seaweed, pebbles, and sand. They have to be tough to survive the ever-changing temperatures and salinity, as well as the predators, that exist in the pools.

Goby fish

Home drilling

Some animals drill into rock to make their homes. A purple sea urchin, left, makes hollows in rock with its strong spines and gnaws rock with its mouthparts. The piddock, right, uses the two halves of its sharp shell to drill into rock. There it lives in safety, breathing and feeding via two siphons that extend from the surface of the rock.

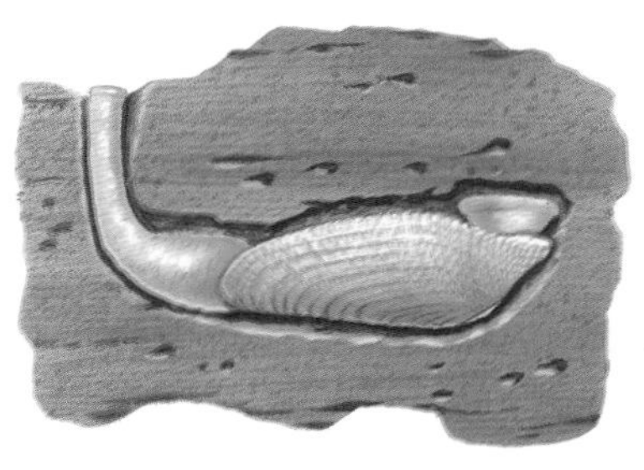

Jagged seaweed
Serrated wrack is brown with branching, leathery fronds that have jagged edges.

Periwinkles
Out of water, periwinkles seal the gap between their shell and rock with mucus.

Common limpet
The conical shape of a limpet's shell helps it resist the pounding of waves.

Feeding prawn
Prawns lurk in tide pools, using their pincers to pick up morsels of food.

Hidden urchin
Some urchins camouflage themselves by holding pebbles, shells, or seaweed over their bodies with their long tube feet.

Breadcrumb sponge
A sponge is a simple animal that encrusts a rock surface so that it is not likely to be washed away by tides.

Cushion star
This starfish eats shellfish, brittlestars, and shore worms.

Snakelock
This anemone is unable to retract its tentacles, and will dry out and die if stranded on a rock.

Limpet

Flat foot

Starfish

Tube foot

Rock grippers

Some animals are well-adapted to clinging to rocks. A limpet has a flat, disk-shaped foot, and clings so firmly that the force needed to move it would break the shell first. Starfish grip rocks with hundreds of tiny suckers called tube feet.

Sheltered housing

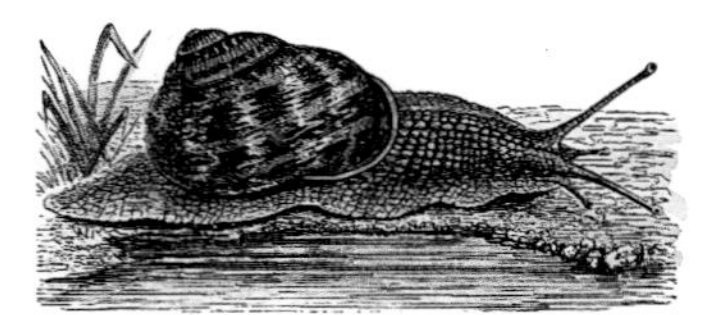

Instead of building their own homes, some small animals take over the already built homes of others. The hard, protective shells of snails and shellfish provide great instant homes. Some animals, such as the pea crab, move in while the original occupants are still alive. Many more creatures use the homes left behind once these occupants have died. This recycling of homes saves the new tenants the time and energy needed to build a home from scratch.

Mobile home

Snails retreat inside their shell homes to avoid very dry or cold weather. They need humid conditions to survive and would soon dry out and die without the protection of a shell.

1 The remarkable female mason bee sometimes lays her eggs inside snail shells for extra protection. The eggs are laid over a few weeks during the spring. A female usually fills several shells with eggs. A large shell may contain up to 20 eggs.

Hollow home
Discarded shell of a brown-lipped snail, found on limestone hills.

Sun shade
Shell is covered with grass to protect the eggs and young from the sun.

Spinning larva
When the larva grows big enough, it spins a cocoon and turns into a pupa.

Solid walls
Walls built by a female mason bee are made of soil and saliva. The walls turn hard when they dry.

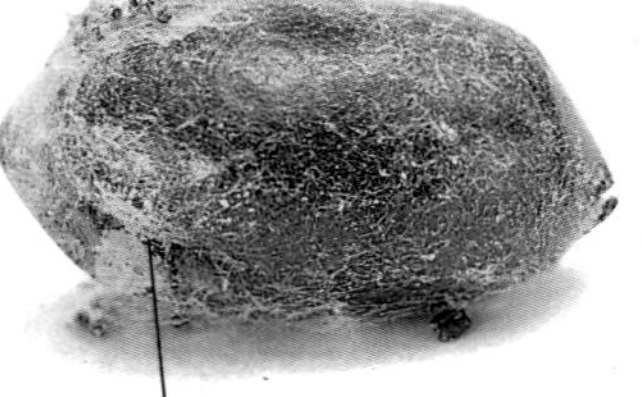

Mason cocoon
Cocoon spun by the larva when it is ready to pupate

2 Inside the shell, the female bee divides the coiled tube into several chambers by building mud walls. She builds one chamber at a time, laying an egg inside and leaving a ball of honey and pollen for the larva to eat when it hatches out of the egg.

Hibernating bee
A fully formed bee has hatched out of the pupa and is waiting for warmer weather before leaving the shell.

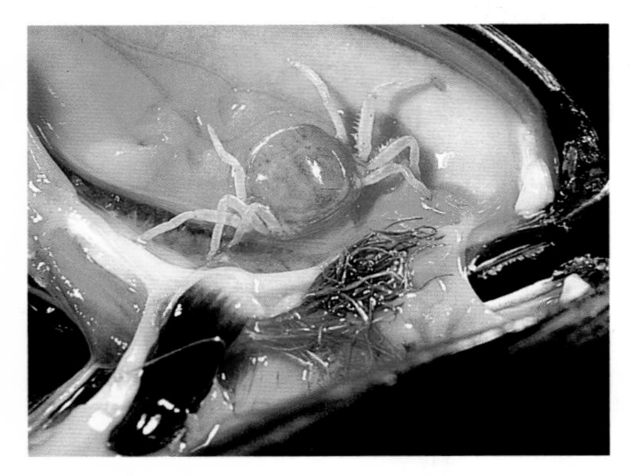

Pea crab

The tiny pea crab takes up residence inside the shells of living shellfish, such as oysters, scallops, and mussels. The shellfish use their gills to trap food particles from the water, so the pea crabs lurk inside the shell, near the gills, catching any food that floats past.

Mimic spider

The mason bee is not the only creature to visit the snail's empty home. This ant-mimic spider has moved into a discarded shell and will probably hibernate there over winter.

New tenant

A hermit crab does not have a hard outer "shell" covering its whole body. The back half of its body is soft and vulnerable to attack. So a hermit crab squeezes its soft abdomen into an empty shell, such as a whelk shell, for protection. As the crab grows, it trades its shell for a larger one. Hermit crabs often share their homes with other creatures, such as anemones or worms.

Busy bee

A female mason bee may have to make between 20 and 30 trips to nearby flowers to collect enough food for each chamber. She collects flower pollen on the hairs under her abdomen and sucks up nectar from the flowers to turn into honey.

Breaking free
This young bee will eventually push through the mud wall and emerge through the hole at the end of the shell.

Larvae nourishment
Mother leaves plenty of food near the egg in each chamber.

3 Young adult bees hatch out of the pupae at the end of summer. They stay in the shell over winter. and break out in the spring. Males emerge first and fly away. By the time females emerge, they have more chance of finding males from a different "home." For the wasps, it is preferable that males and females from the same shell do not mate together.

Animal lodgers

Animals, from tiny dust mites to larger squirrels, live all over our homes.

As more and more of the world becomes covered with towns and cities, some animals have decided to move in with us. They may stay in residence all year round, or visit only in the winter for hibernation. In our homes, animals can take advantage of a warm, sheltered environment and a constant supply of food. The artificial habitats in our homes are similar to many natural ones – house walls are like cliffs, attics are like caves, and wooden furniture is like a fallen log. Some visitors can cause problems, by chewing through wires, spoiling food, or carrying diseases for example, but others are useful because they eat unwanted insects.

Furry visitor

The brush-tailed possum nests in the roofs of buildings in Australia because they are similar to the tree hollows or hollow logs it uses in the wild. It feeds on refuse and garden plants, such as rose buds, grass, and clover. Colonies of these possums live in some parks.

House spider

To a spider, a house is not too different from the caves where its ancestors once lived. There are a few pitfalls however, such as slippery bath and sink surfaces, which a spider's foot cannot grip.

Chimney home
White storks often build their untidy stick nests on the roofs of European houses, instead of on trees or cliffs.

Hanging bats
Bats hang from the steep beams of attics, out of reach of most predators. Young bats can practice flying in the roof space before venturing outside.

Clothes eater

The larvae of clothes moths are able to digest hair, wool, and silk, and so find an abundant food supply in our clothes, carpets, and general household garbage.

Attic nest
Squirrels normally build nests in trees, but a better alternative is a warm, dry attic, out of the wind and rain.

Bloodsuckers
Bedbugs live in birds' nests, animal bedding, and even our own beds. They feed on the blood of mammals.

Under the eaves
Swallows often plaster their mud nests under eaves.

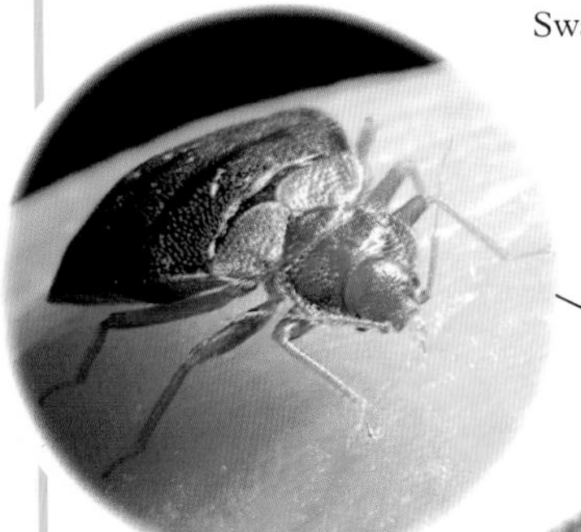

Nesting hole
European starlings may make a safe nest in the corner of an attic.

Bookworm
Book lice live on the pages of books, feeding on tiny molds that grow there. They also feed under wallpaper in damp houses.

Wall climbers
Geckos use house walls as hunting grounds for insects. Ridged toes give them a good grip so they can easily run up and down walls and across ceilings.

Bug in a rug
Carpet beetle larvae are covered with hairs and are often called woolly bears. They cause damage to furs and fabric.

Mouse in the house
A house mouse makes a nest of chewed-up wool, paper, and straw.

Continued on next page

Continued from previous page

A wide diet

Cockroaches like warm kitchens and eat anything, even ink, paint, old shoes, and the bodies of other cockroaches.

Rice eaters

The rice weevil bores through the hard shells of rice grains to eat the insides.

Germ carrier

Common house flies often contaminate human food and spread diseases.

Trash can raiders
In North America, raccoons often raid trash for scraps.

Night hunters
Toads hide in damp places under bricks, stones, and compost heaps.

Pet problems

This cat flea can live on dogs and people as well as cats. It lives on the blood of its host and has sharp mouthparts to pierce flesh.

Winter home

Adult herald moths hibernate in our homes during the winter and emerge when the warmer weather arrives.

Cub nursery
Foxes are happy to raise their cubs in a basement, where it is warm and dry and there is room for the cubs to play.

Sewer dwellers
Rats like to live in the dark sewers under our homes, but may come indoors to search for food.

Invisible invaders

Billions of dust mites live in our beds, carpets, and any dusty corner, munching away on flakes of human skin. This is what makes up most of the dust in our homes. Some people are allergic to dust mites.

Living in the dark

Earwigs hide away in dark, dry crevices during the day, and come out at night to feed on plant or animal food. They hibernate during the winter.

Glossary

A

Abdomen The rear part of the body of an insect or a spider. In vertebrates, the part of the body containing the digestive organs.

Mason bee

Alates These winged termites are the reproductive generation. They fly off to start new colonies when the original colony grows too large.

Allergy A reaction by the body to a foreign substance, such as pollen or dust.

Antennae A pair of jointed feelers on an insect's head that are used for touching and tasting.

B

Bacteria One-celled, microscopic organisms that feed on other organisms, both living and dead.

Bird A warm-blooded, egg-laying animal with wings and a body covering of feathers.

Rotting log

C

Camouflage The method by which an animal hides from its predators by taking on the colors and appearance of its surroundings.

Carnivore A meat-eating animal.

Climate The typical pattern of weather that is experienced in a place over a long period of time.

Cocoon A protective covering for a pupa or the eggs of invertebrates. Some cocoons are made of silk.

Colony A large group of creatures of the same species living together.

Compost Rotting waste vegetation that breaks down, largely as a result of the action of bacteria and fungi, and is used to enrich soil.

Coterie A social group of animals, such as prairie dogs, that defend a common territory against other groups of animals.

Crustaceans Mainly aquatic animals with shell-like bodies, such as crabs, shrimps, or lobsters.

Termite mound

D

Digestion The breakdown of food so that it can be absorbed into the body.

Domestication The taming of animals by people so that they are under human control.

E

Ecosystem All the living and non-living things in a particular area, which link together to form an integrated system. Forests and deserts are ecosystems, but an ecosystem can be much smaller, such as a tree or a drop of water.

Evaporation Changing of a liquid or a solid into a vapor (gas).

F

Fertilizer A material, such as manure, that is used to add nutrients to soil.

Fortress (of a mole) A large mound of soil pushed up by male or female moles in which a female may make a nest and rear her young.

Fungus An organism with no leaves or roots, such as a mold or yeast. Fungi cannot make their own food, so absorb it from living or dead organisms.

G

Gall An abnormal swelling on a plant caused by bacteria, fungi, or some young insects.

Gill Body parts of fish and some aquatic animals that are used for removing oxygen from water to enable them to breathe.

H

Habitat An animal's natural home.

Hibernation The time spent, by some animals, sleeping through the winter months. An animal's body temperature falls and the rate of all its body processes slows down during this time.

Humid A moist atmosphere.

I

Insect A small creature with a body divided into three parts – head, thorax, and abdomen. An insect has three pairs of legs and often has two pairs of wings.

L

Larva (Grub) Immature stage of an animal that is very different in form from the adult, such as a caterpillar.

Sociable weaver bird nest

Under the sand

Life cycle Different stages of development that a living thing passes through from birth until death.

Lodge The home of a beaver, built of sticks and mud in the middle of a dammed river.

M

Mammal A warm-blooded animal with fur or hair on its body. Female mammals produce milk from glands in their bodies to feed to their young.

Metamorphosis The change in form and structure as an animal passes from an embryo to an adult stage of its life cycle. For instance, a butterfly or moth caterpillar changing into a pupa and then into an adult.

Oven bird nest

Molt The periodic shedding of an animal's outer body covering, such as skin, feathers, or hair.

N

Nectar A sugary liquid that is produced in the flowers of some plants to attract insects or birds for pollination.

O

Ovipositor A long, tubelike organ at the end of the abdomen of most female insects. The ovipositor is used for laying eggs.

P

Plankton Tiny animals and plants that drift in salt- or freshwater.

Pollen Fine, powdery substance produced by flowers and used by bees to make honey.

Predators Animals that catch and kill other creatures.

Prey Animals that are hunted and eaten by other animals.

Proboscis The long, tubelike mouthparts of some insects that are used for sucking up food.

Pupa Resting stage in the life cycle of an insect during which the body tissues of the immature insect are reorganized into the adult form.

Q

Queen The female that starts a colony of social insects such as termites or wasps. In many colonies there is only one queen and she is the only member of the colony able to reproduce.

R

Rain forest Dense forests found in hot, humid areas near the equator.

Recycle The reuse of waste materials to save energy and resources.

Reptile A cold-blooded, egg-laying animal with a scaly body covering.

S

Salinity The amount of salt within a solution, such as in seawater.

Saliva Clear fluid produced in the mouth that helps with swallowing and digesting food.

Silkworm cocoons

Savanna This is an area of open grassland with scattered trees and shrubs found in hot, dry areas.

Siphon A tube through which aquatic mollusks draw water in and out of their bodies.

Compost heap

Species A group of animals that are able to breed together.

Spinnerets The organs of spiders and some insects that release the silk made in their silk glands.

T

Tentacles The elongated, flexible structures on the heads of many invertebrates used for feeling, exploring, and grasping things.

Termite An antlike social insect that lives in hot, tropical climates.

Tube feet Outgrowths from the bodies of echinoderms, such as starfish, used for moving, feeding, and breathing.

W

Worker wasp This wasp is responsible for food gathering, nest building, and caring for the young wasps of a colony.

Worker termite An infertile termite that cares for the young, gathers food, and carries out any nest repairs.

Index

A B C

D E F

G H I

J L M

N O P

R S T

V W

Acknowledgments

Design assistance:
Rachael Dyson, Iain Morris, Salesh Patel, and Jason Gonzalez

Additional photography:
Kim Taylor, Jane Taylor, Geoff Dann, Dave King, Peter Chadwick, Colin Keates, Frank Greenaway, Jerry Young, and Richard Davies

Additional models:
Gary Staab

Photoshop retouching:
Bob Warner and Oblong Box

Thanks to:
Dr. Russell-Smith, Natural Resources Institute; Marion Dent for index

Illustrations:
11br, 12m, 13tr, 16mr, 18m, 23tr, 23mr, 24tm, 29mr, 30bl, 31tr, 35tr, Michael Lamb; 11bl Simone End

Picture credits:
r=right, l=left, t=top, c=centre, b=below, a=above
Biofotos: Heather Angel 9tr, 18bl, 26bl, 31tl, tc, Soames Summerhays 30bl; **Bruce Coleman:** Jane Burton 6bc, 10cr, 14br, 25tc, 36bl, P Clement 28tr, Patrick Clement 22bc, 42bl, Jeff Foott Productions 34tr, Clive Hicks 20cl, Udo Hirsh 23cb, 42bc, Stephen J Krasemann 9tc, 40bl, Wayne Lankinen 33bl, Dr Rocco Longo 20cr, 43bc, George McCarthy front cover cb, 15cr, 22bl, WS Paton 8br, Dr Eckart Pott 33cr, 39cr, Alan G Potts 27tr, Dr Sandro Prato 20cla, bl, 20-21, 21cl, 43cb, br, Fritz Prenzel 38tr, Andrew J Purcell 26cl, Andy Purcell 41br, Dr Frieder Sauer 8cl, 21cb, bc, Norbert Schwirtz 16cl, Alan Stillwell 39bl, Jan Taylor 9cr, 29c, Kim Taylor front cover cla, cra, clb, 6tr,14tr, bl, 15tl, tr, 23cl, cb, ca, 25bc, br, 38tc, 39br, cl, 41cl, 42clb, c, cb, Peter Ward 10cl, Gunter Ziesler 19tl; **Mary Evans Picture Library:** 24tl; **Hutchison Library:** John Wright 10bl; **Image Select /Ann Ronan:** 10tl, 20tl, 22tl; **Microscopix:** Andrew Syred 41bl; **Natural History Photographic Agency:** Anthony Bannister 12bl, 28c, Stephen Dalton 6cb, 25cr, Nigel Dennis 18 crb, Ron Fotheringham 24cr, 43cr, EA Janes 25cl, Peter Pickford 19cb, 42br; **Oxford Scientific Films:** Bob Bennett 33crb, Mike Birkhead 39bl, Deni Brown 22cl, MJ Coe 9br, JAL Cooke 23tr, crb, 23br; John A Cooke 29t, Bruce Davidson 10tr, David G Fox 28bl, Max Gibbs 31br, Mark Hamblin 38cla, Lon E Lauber 33tc, Will Long & Richard Davies 26-27b, Steve Turner 10cbl; /Animals Animals: Dr Mark Chappell 33tr; /Survival Anglia: Alan Root 12bc, br, 13cb **Planet Earth Pictures:** John Downer 39tl, Peter Gasson 32tr, Steve Hopkin 40cl; **Premaphotos Wildlife:** KG Preston-Mafham 21crb; **Science Pictures Limited:** 21tr; **Still Pictures:** Klein/Hubert 16br; **Tony Stone Images:** Paul Chesley 20clb; **Kim Taylor:** 6cb, 24c, 25tr, 37r, 43ca; **Telegraph Colour Library / Colorific!:** David Young 27cl; **Trip:** W Jacobs 33tl; **Zefa Pictures:** 16bl, 19tr; / Minden: Jim Brandenburg 32b

Every effort has been made to trace the copyright holders. Dorling Kindersley apologizes for any unintentional omissions and would be pleased, in such cases, to add an acknowledgment in future editions.